Read for a Better World™

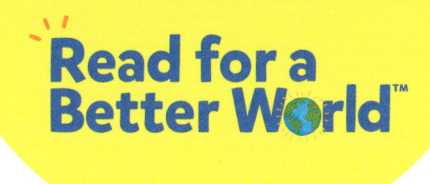

BRACHIOSAURUS
A First Look

JERI RANCH

GRL Consultant, Diane Craig, Certified Literacy Specialist

Lerner Publications ◆ Minneapolis

Educator Toolbox

Reading books is a great way for kids to express what they're interested in. Before reading this title, ask the reader these questions:

What do you think this book is about? Look at the cover for clues.

What do you already know about this dinosaur?

What do you want to learn about this dinosaur?

Let's Read Together

Encourage the reader to use the pictures to understand the text.

Point out when the reader successfully sounds out a word.

Praise the reader for recognizing sight words such as *had* and *were*.

TABLE OF CONTENTS

Brachiosaurus 4

You Connect! 21
STEM Snapshot 22
Photo Glossary 23
Learn More 23
Index. 24

Brachiosaurus

Brachiosaurus is a kind of dinosaur. It lived 150 million years ago.

Brachiosaurus
brack-ee-uh-SOAR-us

The dinosaur was very big. It was almost forty feet tall.

How tall are you?

8

It was as heavy as four elephants. It was as long as two buses.

elephant

It had a long neck.
Its tail was shorter.

The dinosaur had tall front legs.

neck

tail

front legs

12

Its teeth were round.
They were shaped
like spoons.

The dinosaur ate leaves. Its neck helped it reach them.

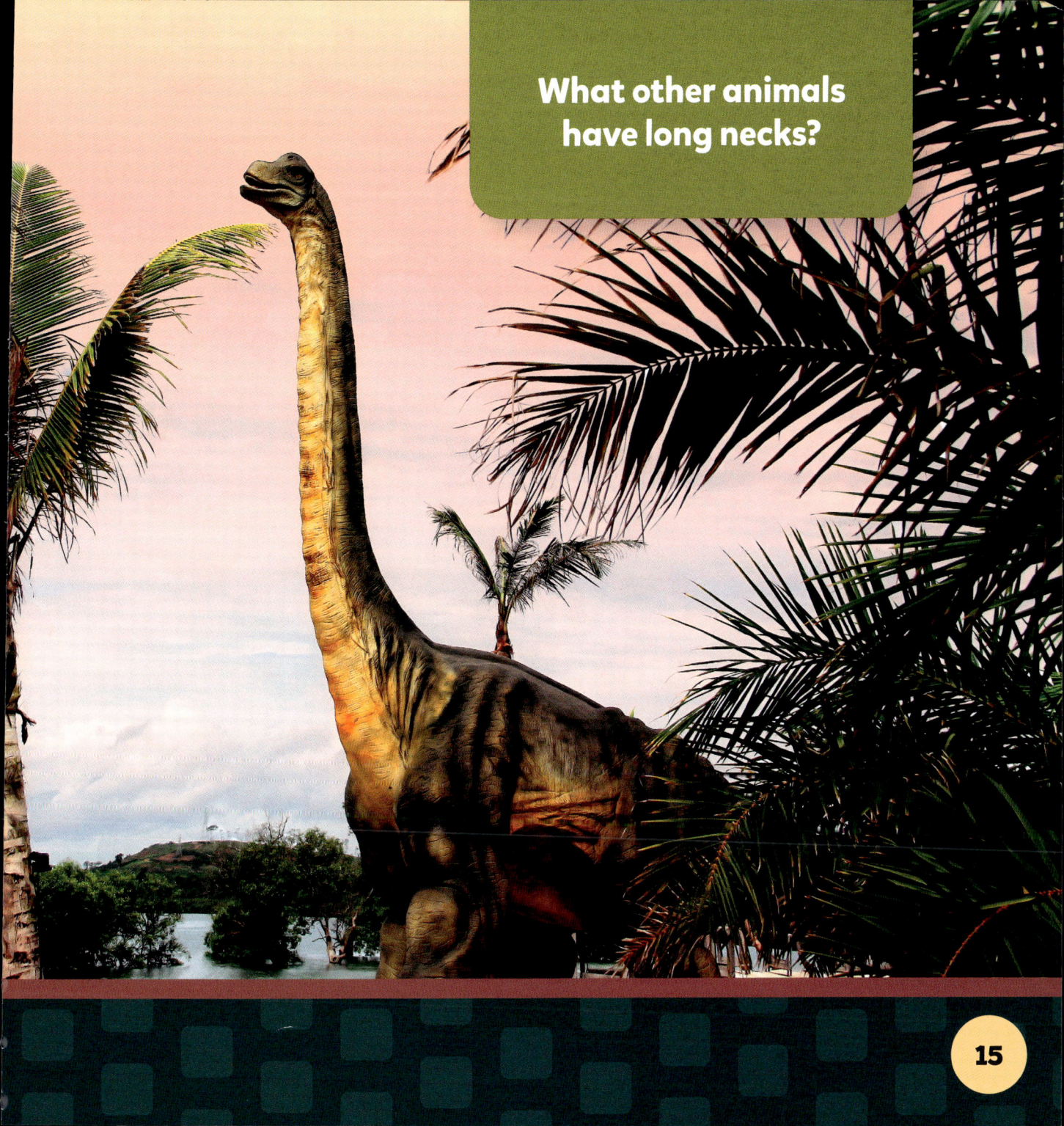

What other animals have long necks?

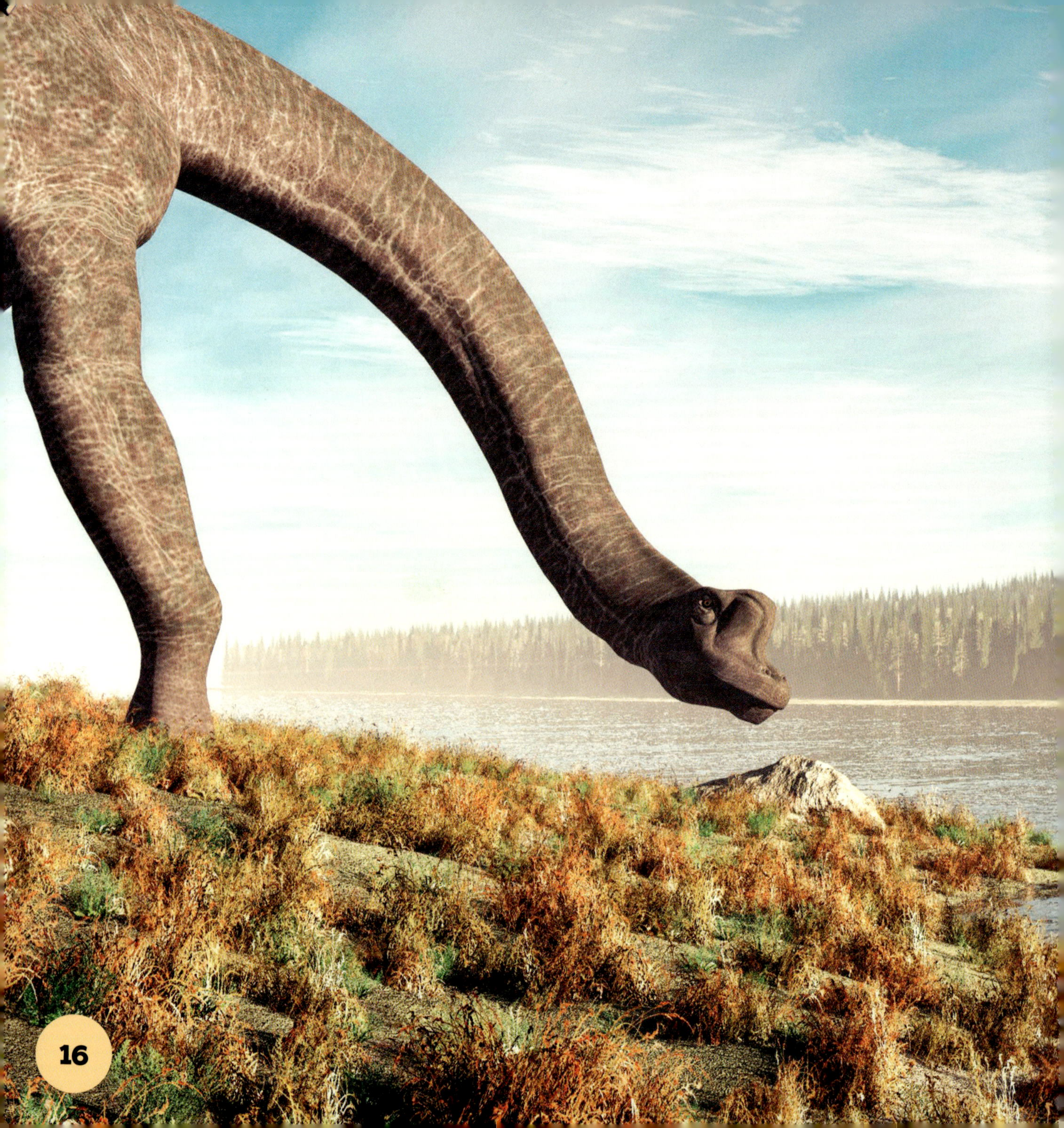

It also ate plants on the ground.

17

This dinosaur is not alive today. But people find its bones.

Would you want to find dinosaur bones?

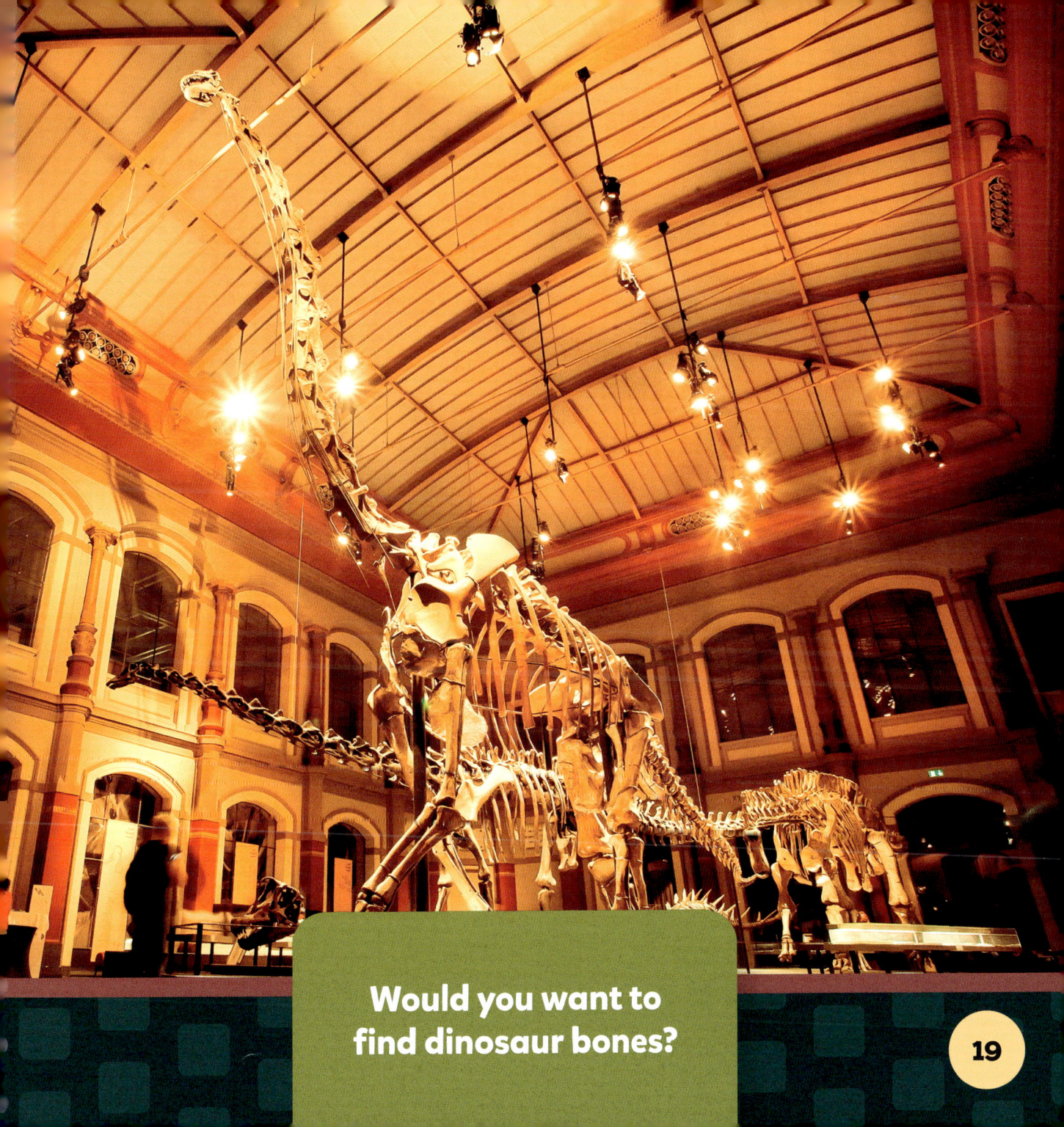

19

Bones teach us about this dinosaur.

You Connect!

What is something you like about this dinosaur?

What else is as big as this dinosaur?

What other dinosaurs do you know about?

STEM Snapshot

Encourage students to think and ask questions like scientists. Ask the reader:

What is something you learned about this dinosaur?

What is something you noticed in the pictures of the dinosaur?

What is something you still don't know about this dinosaur?

Photo Glossary

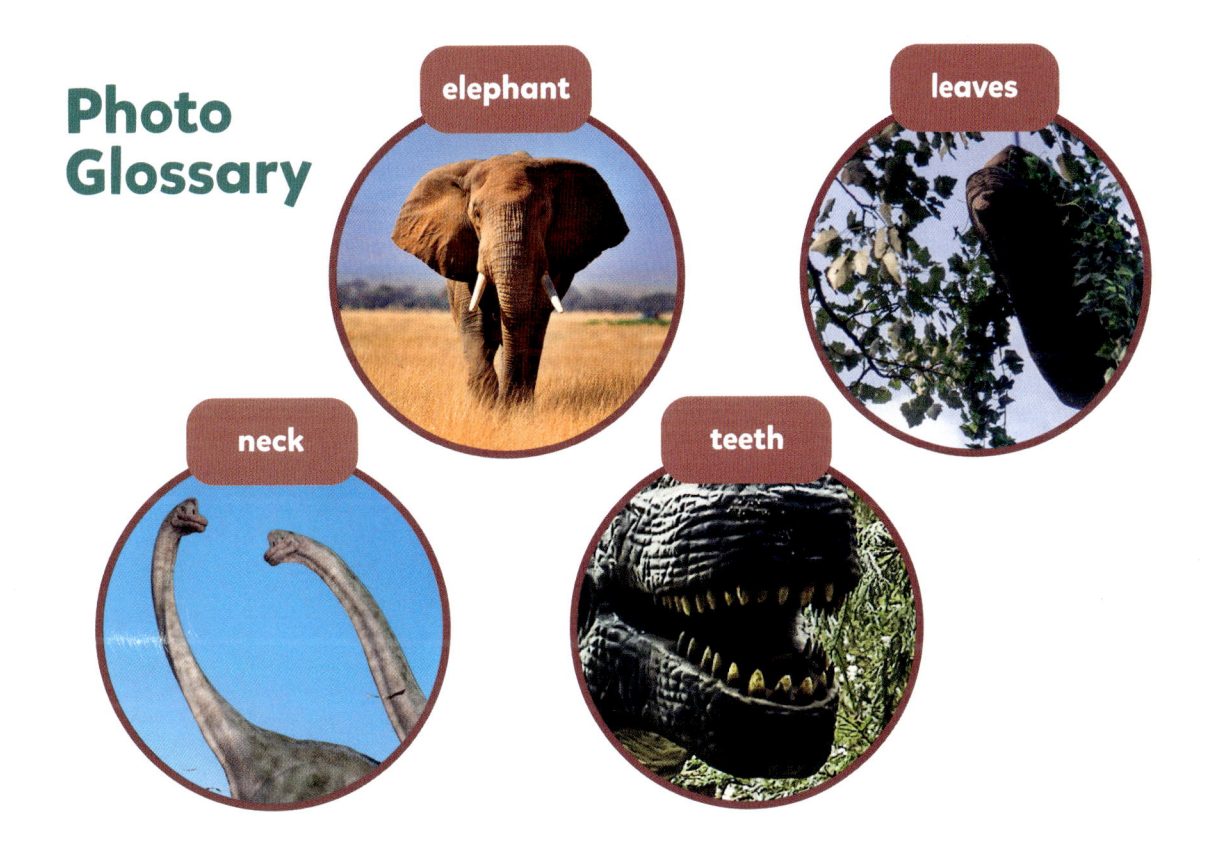

elephant

leaves

neck

teeth

Learn More

McDonald, Jill. *Exploring Dinosaurs*. New York: Doubleday Books for Young Readers, 2023.

Nelson, Jake. *I'm a Brachiosaurus*. Ann Arbor, MI: Cherry Lake Publishing, 2021.

Sabelko, Rebecca. *Brachiosaurus*. Minneapolis: Bellwether Media, 2022.

Index

bones, 18, 19, 20
leaves, 14
legs, 11

neck, 10, 14, 15
plants, 17

teeth, 13

Photo Acknowledgments

The images in this book are used with the permission of: © Orla/iStockphoto, pp. 4–5; © Dotted Yeti/Shutterstock Images, pp. 6–7; © Alberto Andrei Rosu/Shutterstock Images, pp. 8–9; © Pooja Prasanth/Shutterstock Images, pp. 9, 23 (elephant); © dottedhippo/iStockphoto, pp. 10, 23 (neck); © YuRi Photolife/Shutterstock Images, p. 11; © AKKHARAT JARUSILAWONG/Shutterstock Images, pp. 12–13; © Maksim Shchur/iStockphoto, pp. 13, 23 (teeth); © Isaac74/iStockphoto, pp. 14, 23 (leaves); © thaloengsak/Shutterstock Images, pp. 14–15; © ALLVISIONN/iStockphoto, pp. 16–17, 20; © Radiokafka/Shutterstock Images, pp. 18–19.

Cover Photograph: © dottedhippo/iStockphoto

Design Elements: © Mighty Media, Inc.

Copyright © 2024 by Lerner Publishing Group, Inc.

All rights reserved. International copyright secured. No part of this book may be reproduced, stored in a retrieval system, or transmitted in any form or by any means—electronic, mechanical, photocopying, recording, or otherwise—without the prior written permission of Lerner Publishing Group, Inc., except for the inclusion of brief quotations in an acknowledged review.

Lerner Publications Company
An imprint of Lerner Publishing Group, Inc.
241 First Avenue North
Minneapolis, MN 55401 USA

For reading levels and more information, look up this title at www.lernerbooks.com.

Main body text set in Mikado a Medium.
Typeface provided by Hannes von Doehren.

Library of Congress Cataloging-in-Publication Data

Names: Ranch, Jeri, author.
Title: Brachiosaurus : a first look / by Jeri Ranch.
Description: Minneapolis : Lerner Publications , [2024] | Series: Read about dinosaurs (Read for a better world) | Includes bibliographical references and index. | Audience: Ages 5–8 | Audience: Grades K–1 | Summary: "The Brachiosaurus will impress readers with its long features and enormous size. Photorealistic images and leveled text will help readers picture just how massive this dinosaur used to be"– Provided by publisher.
Identifiers: LCCN 2022039321 (print) | LCCN 2022039322 (ebook) | ISBN 9781728491325 (library binding) | ISBN 9798765603468 (paperback) | ISBN 9781728499185 (ebook)
Subjects: LCSH: Brachiosaurus–Juvenile literature.
Classification: LCC QE862.S3 R3585 2024 (print) | LCC QE862.S3 (ebook) | DDC 567.913–dc23/eng/20220826

LC record available at https://lccn.loc.gov/2022039321
LC ebook record available at https://lccn.loc.gov/2022039322

Manufactured in the United States of America
1 – CG – 7/15/23